ENERGY IN THE VEDAS

Dr. Ravi Prakash Arya

Amazon Books, USA

in Association with

Indian Foundation for Vedic Science

H.O.1051, Sector-1, Rohtak, Haryana, India

Ph. 9313033917; 9650183260

Email: vedicscience@gmail.com

vedicscience@rediffmail.com

First Edition

Kali era 5117 (c. 2014)
Kalpa era 1,97,29,49,117
Brahma era 15,55,21,97,29,49,117

Second Edition

Kali era 5123 (c. 2021)
Kalpa era 1,97,29,49,123
Brahma era 15,55,21,97,29,49,123

ISBN 81- 87710-79-9

Contents

INTRODUCTION

The Vedas describe energy by the name of agni. The same agni, when located in various spaces, is named variously as Indra, Vāyu, Jātavedas, Vaiśvānara, Puriṣya, Śuci, Pavamāna, Pāvaka, Aśva, Gau, Ajā, Avi, etc. Some 200 Sūktas have been devoted to Agni (energy) in Ṛgveda. Agni is noted to devour his parents (matter) soon after its birth. This points to nothing else but the conversion of matter into energy.

The *Aiteraya Brāhmaṇa* (2.3) and the *Taittirīya Br.* (1.4.4.10) upholds that all heavenly bodies in the universe originated from Agni (energy). The *Śatapatha Brāhmaṇa* (1.6.2.8), the *Tāṇḍya Brāhmaṇa* (9.4.5.18), Ṣaḍviñśa Brāhmaṇa (3.7) and The *Gopatha Brāhmaṇa* (Second part, 1.12.-13) also observe in a similar way. According to them, energy is behind the birth of the entire material world. The *Śatapatha Brāhmaṇa* (6.1.2.8) further says that energy involves entire material creation. At another place, it is observed that one and the same agni has transformed into various material objects. According to the *Aitareya Brāhmaṇa* (2.6), all the matter particles (material objects) in the observer space are made up of agni (energy). The *Aitareya Brāhmaṇa* uses the word paśu to denote matter particles. The *Taittirīya Brāhmaṇa*, (1.1.4.3) also observes that all matter particles are born of agni or energy. The *Kapiṣṭhala Saṁhitā* (31.19) also says that matter particles are created from agni (energy). According to the *Śatapatha Brāhmaṇa* (6.1.4.12), all matter particles are transformed from Agni. In view of the foregoing discussion, it may be ascertained that this material creation is the result of the transformation of energy into matter. Or one can say, change of energy into matter causes the origin of this material world and inversely change of matter into energy leads to the decreation or devolution of the world.

The present paper tries to discover the concept and various forms of energy known to the Vedic people. It also discloses the real meanings of Puruṣa medha, Aśva medha, Gau medha, Ajāmedha and Aśvamedha as intended by the Vedic seers in context of energy. The solar and geothermal energy was allowed to be harnessed for various useful applications. The very idea that life on earth is sustained by the sun alone is not acceptable to Vedic seers. According to them, the earth is equally responsible for the sustenance of life on it. The planets without their own energy are unable to sustain life even in the presence of the sun. It is the permutation and combination of solar radiation and geothermal energy that generates the atmosphere conducive to sustain bio-life on the earth.

The energy generated from fossil fuels was called

Purisa Agni (excrement or faeces of the earth). Vedas prohibit the use of fossil fuel energy and stressed the need to harness the solar and geothermal energy for various use by earthlings, as they were very conscious about the healthy and clean environment and never allowed any act that may lead to pollution of the atmosphere.

For earthlings there are two basic sources of energy: firstly, the Self Sustained Natural Nuclear Fission Reactor operating in the core of the Earth and secondly, the Self-sustained Natural Nuclear Fusion Reactor operating in the core of the Sun.

The paper also discusses three Sanskāras of energy as discovered by the Vedic seers.

1. Sthiti-sthāpaka sanskāra (potential) emerging from Tamoguṇa, a quality of Prakṛti (matter).

2. Vega (kinetic) emerging from Rajoguṇa, a quality of Prakṛti (matter) and

3. Bhāvanā sanskāra (psychic energy) emerging from Sattvaguṇa, a quality of Puruṣa (self).

1

THREE SANSKĀRĀS OF ENERGY

Modern science considers the world of animate or inanimate things, as the composition of matter. As such energy, in modern science, is described as mechanical in nature which can further be divided into two parts, viz. potential and kinetic. Potential energy is described as the energy stored in a body due to its position, whereas, kinetic energy is defined as the energy possessed by a body due to its motion.

On the other hand, the Vedic scientists did not consider the world as merely the composition of matter. According to them, the world is not the composition of Prakṛti (matter) alone, rather it is composed of both Prakṛti (matter) and *Puruṣa* (consciousness). That is why, the Vedic scientists considered this creation as a composition of triguṇas (*sattva*, *rajas* and *tamas*)- *sattva* belonging to Puruṣa and *rajas* and *tamas* belonging to Prakriti or matter. The result is that the Vedic concept of energy is also broader than that of the modern science. Vedic scientists represented by Praśastapāda (6.10, Sanskāra Prakaraṇa) in his famous work 'Padārthadharmasaṅgraha' which is considered to be a commentary on the *Vaiśeṣika Darśana*, and describes energy both mechanical as well as psychic.

According to him,

संस्कारस्त्रिविधः उक्तः वेगोभावनास्थितिस्थापकश्च।

There are three types of sanskāra. 1. Vega Sanskāra (kinetic energy), 2. Bhāvanā Sanśkāra (psychic energy), and 3. Sthisthāpaka Sanśkāra (potential energy).

Mechanical energy is qualified by two SaŠskāras (1) Sthitisthāpaka and (2) Vega and psychic energy is qualified by Bhāvanā Sanskāra. Sthitisthāpaka Saṁskāra is akin to that of potential energy which is the result of tamoguṇa, whereas Vega Saṁskāra is similar to that of kinetic energy which is the result of rajoguṇa. The third Bhāvanā Sanskāra or the psychic energy is the result of sattvaguṇa. This energy is quite new to the modern science. If energy is defined in terms of the ability to do work, Bhāvanā Sanskāra or psychic energy will provide unlimited ability to do work. Through one's mental power acquired through the regular practice of Yoga, one can do any work. He can send messages to any part of the globe or to other planets if need be. He can read even other person's minds and is able to materialise anything at his will.

Here it would also be worth mentioning that Sanskāra has been enumerated in the *Vaiśeṣika Darśana* among the Guṇas or qualities. Hence, it is easy to infer that Vedic scientists never took energy as an independent entity as is considered in the modern science. Guṇa or attribute always resides in Guṇī, subject (attributant), the object. Shakti (power) resides in shaktimāna (powerful). For example, the mechanical energy- potential or kinetic- cannot exist independent of a body due to its position or due to its motion. Similarly, psychic energy cannot exist independently of Consciousness, i.e God. Thus Vedic philosophers never think of creation without its creator.

2

PURUṢA: THE MASS-ENERGY OF UNIVERSE

In the Vedas, the Universal mass-energy is called Puruṣa which is the life principle of the Universe. The *Chāndogya Upaniṣad*, compares Puruṣa with the universe. The *Śatapatha Brāhmaṇa* (13.6.2.21) also observes in the same manner as:

ime vai lokāḥ pūḥ, ayam eva puruṣo yo'pavate. so'syāṁ puri śete tasmāt puruṣaḥ. yadeṣu lokeṣvannaṁ tadasyānnaṁ medhaḥ. tad-yad asy-aitad annaŠ medhas tasmāt puruṣmedhaḥ. atho yadasmin medhyān puruṣā nālabhate tasmād eva puruṣa-medhaḥ

[Meaning] This universe is like the physical body. One who resides in this body is called Puruṣa. Mass energy in the universe acts as food since this food is consumed for the evolution of the universe. That is why, the Puruṣamedha (evolution of the universe) takes place.

This puruṣmedha works on the principle of one particle combining with another particle to generate a third particle and so on.

The *Charaka Saṁhitā* (*Śārīra-sthāna*, 5.3),

describes Puruṣa an epitome of the Universe. The concept of 'Brahmāṇḍa Puruṣa' is the outcome of the above Vedic concept which establishes a parallelism between the organic whole and the universal whole.

According to the *Puruṣa Śukta* of the *Ṛgveda* (10.90.10), the Puruṣa (entire mass-energy of universe) gives birth to four paśus: avi, aśva, ajā and gau in observer space. The mantra says, "From that (Puruṣa) horses were born, who have teeth on both sides. From that cows were born, from that goats were born and sheep were born."

Thus observer space (universe) became conspicuous with the presence of four grāmya paśus like aśva, gau, avi and ajā. Here aśva represents stars, gau symbolises planets and satellites, avi represents intermediate space and aja is the black holes existing in the universe. These all-material bodies are the source of energy. For example, stars are the source of solar energy, planets are the source of geothermal energy and intermediate space is the source of field energy. This fact has been corroborated by Śatapatha Brāhmaṇa (6.2.1.-4) as: Prajāpati saw agni in those paśus, therefore they are called paśus. (Note: The word paśu is derived from root dṛś (paśya) 'to see'.)

The *Yajurveda* (23.17) describes Agni as 'paśu'. The *Taittīriya Brāhmaṇa* (1.1.4.5) mentions paśus as āgneya'. All these Vedic authorities confirm that the objects of observer space (universe) are made up of energy and they generate energy.

Hereunder, we shall take stock of the four forms of energy- aśva (horse), gau (cow), ajā (goat) and avi (sheep)-existing in the Universe, according to the Vedic seers.

3

AŚVA: THE SOLAR ENERGY

The Aśva in the Vedas is described as one among the 4 grāmya paśus, already mentioned above. Aśva here is nothing but the sun which is the biggest source of energy in our universe. Ṛgveda (1.163.2) compares the sun with aśva (horse). Taittirīya Brāhmaṇa (3.9.23.2) describes aśva as āditya. Aitareya Brāhmaṇa (6.35) more emphatically mentions the radiant sun as white aśva - *atha yo'sau (suryaḥ) tapatī eṣo'śvaḥ śveto rūpaŠ kṛtvā'śvābhidhānyapihitenātmanā praticakrāma.* Gopatha Brāhmaṇa (Second part 3.19) also calls sun as aśva - sauryyo vā aśvaḥ [Sun is verily aśva]. In Vedic and Paurāṇika allegories, the sun has been described as a chariot yoked with seven horses. These seven horses of the sun are nothing but the seven- vibgyor- rays of the sun.

In Yajurveda, (23.53), the following query has been raised.

> *kā svid āsīt pūrva-cittiḥ. kiṁ svid āsīd bṛhad vayaḥ. kā svid āsīt pilippilā. kā svid āsīt piśiṅgilā.*

> [Meaning] What is the first storehouse (of energy)? What is the biggest source of energy in

the observer space? What object is pilippilā (protector) and what object is piśaṅgila (devourer)?'

The answer given in the next mantra (Yajurveda, 23.54) is as under:

dyaur āsīt pūrvacittir aśva āsīd bṛhad vayaḥ.
avirāsīt pilippilā, rātrir āsīt piśaṅgilā.

[Meaning] Light space (source of primary and pure energy) is the first storehouse of energy. Aśva (stars/sun) is the biggest source of energy in the observer space. Avi (intermediate space or magnetosphere) is the protector and ajā (black hole) is the devourer of everything.

4

AŚVAMEDHA YĀGA: A PROCESS OF HARNESSING SOLAR ENERGY

The word Aśvamedha is formed of two words, Aśva + medha. The meaning of aśva has already been elaborated as the sun. Medha is derived from root medhṛ 'to achieve' or 'to kill'. The process of medha involves the gain of matter and loss of energy and vice versa. Aśvamedha yāga is nothing else but a process to harness solar energy for the sustenance of life on the earth. The Ṛgveda (2.167.1), speaks about harnessing the energy from the sun for useful technical applications. It also highlights earth receiving energy from the sun as:

yamena dattaṁ trita enam āyunag indraŠ prathamo adhyatiṣṭhat

gandharvo asya raśanām agṛbhṇāt sūrād aśvaŠ vasavo niratasṭa.

[Meaning] (vasavaḥ) The scholars (niratasṭa) harness (aśvaŠ) energy from (sūrād) sun (yama) in a controlled manner (ayunak) and utilized for various technological purposes. The energy was first transformed into Indra (electricity) for its applied use. Gandharva (magnetosphere of the earth) (agṛbhṇāt) captured the (raśanām) reins (radiations) from (sūrād) the sun.

Śatapatha Brahmaṇa (9.4.2.18) says that Aśvamedha is the sun - *asāvā'ditya'śvamedhaḥ*. At another place Śatapatha Brahmaṇa (10.6.5.8) says that Aśvamedha is performed by the radiation heating from the sun- *eṣa vā aśvamedho ya eṣa (sūryaḥ) tapati*. Śatapatha Brahmaṇa (11.2.5.4) also says that Aśvamedha is to be performed year after year.

Following rituals are involved in the performance of Aśvamedha yāga.

4.1 Year long ritual

Aśvamedha yāga is performed for a duration of one year, which symbolizes Earth's revolution around the sun.

4.2 Emperor (Cakravrati Samrāṭ)

Aśvamedha yāga can be performed only by an Emperor (Cakravrati Samrāṭ). Here Cakravrati Samrāṭ is symbolic of the sun, because the sun alone is the emperor of the Solar system.

4.3 Requirement of Horse

Such a horse is required for the ritual whose forepart is black and back part is white which has a cart-shaped mark on its forehead. The horse of Aśvamedha yāga with the above features is symbolic of Sun since out of 24 hours of a solar day, the first 12 hours are covered by night and the second 12 hours are covered by day. The night is represented by a dark hue and day by white. The cart sign on the forehead of the horse of Aśvamedha yāga represents the twilight hours when the rays of the rising sun in the eastern horizon give an impression of a cart.

4.4 Rein of Horse

Aśvamedha yāga horse is supposed to have a rein measuring 12 to 13 aratnis (units). This measurement of 12 to 13 aratnis symbolizes 12 months or 13 months (in the case of an intercalary month) of a year. Clarified butter is applied to the rein of the horse, which symbolizes the luminosity of the sun.

4.5 Four queens of the Emperor

The Aśvamedha yāga is performed by a consecrated king, who is accompanied by four queens. Here four queens of the king are symbolic of four directions. Mahiṣī (queen dowager) symbolises Eastern direction. The Sun rises in this direction. That is why it is figuratively said in Śatapatha Brahmaṇa (13.5.2.2), that the phallus of a horse is placed in the lap of Mahiṣī queen-*nirāyatyāśvasya śiśnaṁ mahiṣyupasthe nidhatte vṛṣā vājī retodhā reto dadhātviti.* Queen Vallabhā or Vāvātā (favourite) is symbolic of the western direction. As the sun sets in the west, so it is allegorically mentioned that the sun takes a rest or sleeps in the western direction. Due to this reason, the king performing Aśvamedha yāga is advised to take a nap resting his head in the laps of Vallbhā or

Vāvātā queen. Similarly, queens named Avallbhā or Parivṛktā (unfavourite) and Dūtaputrī or Pālāgalī (daughter of envoy) represent north and south directions respectively because these directions can have their contact with the sun only during Uttarāyaṇa (winter solstice) and Dakṣiṇāyana (summer solstice).

4.6 Horse's year-long wandering

The year-long wandering of the horse is symbolic of the earth's period of revolution around the sun.

4.7 Bodyguards of the horse

Sun rays are the representative of bodyguards of the horse. The *Ṛgveda* (6.47.18) mentions tens of hundreds of rays of the sun -*yuktā haryaś śatādaśa.*

4.8 Fastening of horse with ropes

In the Aśvamedha yāga, the horse is fastened with ropes from all sides. This is symbolic of the sun being surrounded by rays. Some other animals are also tied to the rope surrounding the horse which symbolises the planets attracted to the sun due to gravitational pull.

Thus it is proved that Aśvamedha yāga is nothing else but a process to harness solar energy for the sustenance of life on the earth. The Ṛgveda (2.167.1), speaks about harnessing the energy from the sun for useful technical applications. It also highlights the earth receiving energy from the sun.

yamena dattaṁ trita enam āyunag indraṁ prathamo adhyatiṣṭhat gandharvo asya raśanām agṛbhṇāt sūrād aśvaŠ vasavo niratașța.

[Meaning] The scholars (vasavaḥ) harness energy from the sun in a controlled manner (yama) and utilized (ayunak) for various technological purposes. The energy was first transformed into Indra (electricity) for its applied use. Gandharva (magnetosphere of the earth) captured the reins (radiations) from the sun.

The *RV.* (3.2.3) talks about the multiple technologies developed from solar energy for long-term benefits, as

kratvā dakṣasya taruṣo vidharmaṇi devāso agniṁ janayanta cittibhiḥ

rurucānaṁ bhānunā jyotiṣā mahāmatyaṁ na vājaṁ saniṣyannupa bruve

[Meaning] The scholars (endowed) with intelligence, harness solar energy powerful like stead for multiple technological uses in order to get efficiency in accomplishing their tasks.

The mantra clearly points out the development of affordable, inexhaustible and clean solar energy technologies which will have huge long-term benefits. It will increase the countries' energy security through reliance on an indigenous, inexhaustible and mostly import-independent resource, enhance sustainability, reduce pollution, lower the cost of mitigating climate change.

As per modern estimates, the total solar energy

absorbed by Earth's atmosphere, oceans and landmasses is approximately 3,850,000 exajoules (EJ) per year. In 2002, this was more energy in one hour than the world used in one year. Photosynthesis captures approximately 3,000 EJ per year in biomass. The amount of solar energy reaching the surface of the planet is so vast that in one year it is about twice as much as will ever be obtained from all of the Earth's non-renewable resources of coal, oil, natural gas and mined uranium combined. The scripture says- "*Sūrya ātmā jagatasthuśaś ca*" [The Sun is the soul of this world, animate and inanimate].

5

GENERATION OF SOLAR ENERGY

The *Ṛgveda* (5.62.1-2) sheds ample good light on the generation of solar energy (radiation + light). Herein it is stated that only one fundamental particle named proton (positively charged matter particle) plays a catalyst role in the continuous generation of energy from the sun. The *mantra* reads as follows:

ṛtena ṛtam apihitaṁ dhruvaṁ vāṁ sūryasya
yatra vimucantyaśvān

daśa śatā saha tasthustadekaṁ devānāṁ
śreṣṭhaṁ vapuṣām apaśya.

[Meaning] O Mitra and Varuna! Your eternal nature is concealed in the permanent flow of thousand rays of sun abiding together. I have detected one of you i.e positively charged particles in the sun rays.

The deity of the above mantra is Mitra and Varuṇa. Varuṇa is the representative of the electron (negatively charged particles) and Mitra is the proton (positively charged particles). The seer wants to say that the sun rays are the abode of protons which are the catalyst to solar energy (radiation+light).

tatsu vāṁ mitrāvaruṇā mahitvam īrmā
tasthuṣīrahabhir duduhre

viśvāḥ pinvathaḥ svasarasya dhenā anu
vāmekaḥ pavirā vavarta.

[Meaning] Exceeding is that your greatness, O Mitra (proton) and Varuṇa (electron), the permanent catalyst of solar energy (radiation+light). Both of you are ever-moving and augment all the self-moving rays of the sun. One of you (proton) perpetually makes the wheel of the sun move.

In the above mantra, it is observed that:

1. Sun has a catalyst.

2. One (Mitra) of two (Mitra and Varuna) is responsible for rotating the wheel of the sun.

3. In the process, the sun's rays are augmented.

4. The flow of rays is caused by the density of radiation.

5. The rays of the sun are self-moving and radiation carries energy.

6

AVI: THE FIELD ENERGY OR THE ENERGY OF INTERMEDIATE SPACE

The Avi is described in the Vedas as intermediate space between stars and their planets or say between our sun and earth. In the *Yajurveda* (23.54), Avi is described as pilippilā which is something very soft that protects and presses very easily. This soft object is nothing else but the intermediate space or magnetosphere of the earth having dense field lines. The magnetosphere of earth protects us from the ultraviolet radiation from the sun. Wool of avi (sheep) is used to strain soma juice. The intermediate space acts as a strainer of soma juice, the radiation from the sun filters down to earth through the magnetosphere of the earth (interface between sun and earth).

Agni is the first form of energy abiding in light space. The second form of energy is described as 'vāyu', filed energy located in the intermediate space or in the magnetosphere of the earth. The *Taittirīya Upaniṣad* describes the constitution of adhiloka as follows: adhiloka is constituted with the earth as the first constituent, sun as the second constituent, antrikṣan (intermediate space) as a link between the two, i.e. earth and sun. Thus vāyu here is not the 'air' but the field energy or magnetic field of the Earth. The 'vāyu' (filed lines or field energy) is a link maker between the two i.e. earth and sun.

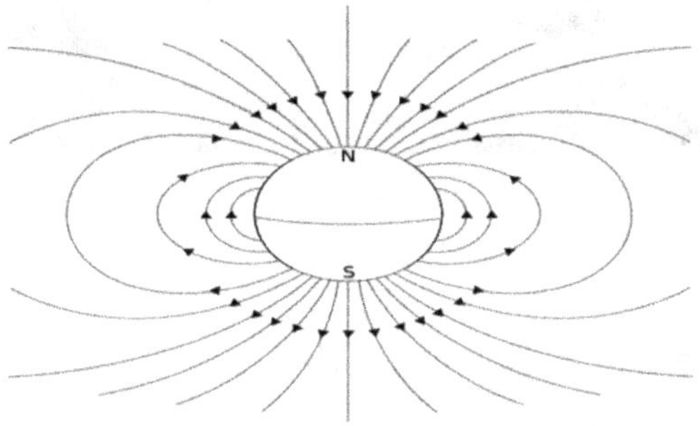

The magnetic field of the earth

Figure-1

In fact, the magnetic field of the earth forms the magnetosphere, which deflects particles from the Solar wind.

The magnetosphere shields the surface of the Earth from the charged particles of the solar wind. The magnetosphere of each and every heavenly body is generated by the plasma located as the nuclear reactor in their centre (wombs). Similarly, the magnetosphere of the earth is generated by the plasma located as the nuclear reactor in the centre of the earth. It is compressed on the day (Sun) side due to the force of the arriving particles and extended on the night side.

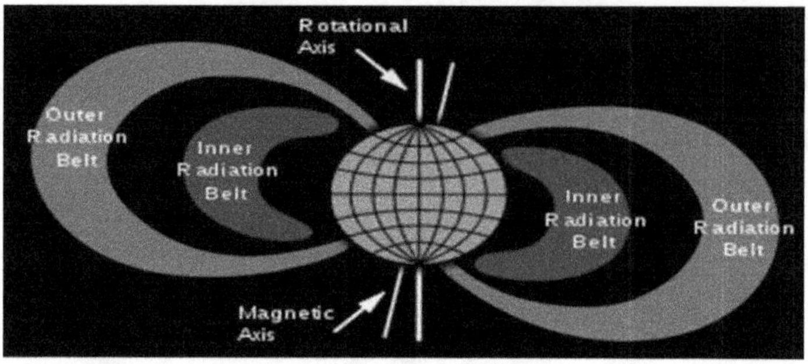

Radiation belts

Figure-2

The collision between the magnetic field and the solar wind forms the Radiation belt, a pair of concentric, Torus-shaped regions of energetic charged particles. These radiation belts surrounding the earth are known in modern science as Van Allen belts. Similar radiation belts are also formed around other planets.

Thus if avi is the intermediate space, Avi medha is the process of collision between the magnetic field and the solar wind forming the Radiation belt. It has been discovered that the Earth's atmosphere limits the belts' particles to regions above 200 - 1,000 Kilometre. Here it is significant to understand that until such times when the magnetosphere of the earth was not formed, the earth was a barren planet. Nothing could grow over it. As soon as the avi (magnetosphere) was formed, the earth became laden with vegetations.

This fact is very beautifully depicted in the *Taittirīya Saṁhitā* (2.1.2.3) as under:

sā avir vaśā samabhavat. te deva abruvan evapaśur vā ayaṁ samabhūt. kasmā imam ālapsyāmaha iti. Atha vai tarhyalpā pṛthvyāsīt. ajātā oṣadhayaḥ. tām aviṁ vaśām ādityebhyaḥ kāmāyālabhanta, tato vāaprayat pṛthivi, ajāyanta oṣadhayaḥ.

[Meaning] Then appeared avi (magnetosphere of the earth). The gods proclaimed that this mass-energy of field has originated. What should we harness it for? By that time earth was a barren planet. No vegetation had grown on it. That avi was harnessed for life on the earth. Thence the earth became fertile and vegetations grew on it.

The similar observation is attested in the *Maitrāyaṇī Saṁhitā* as:

athavā iyaṁ tarhyukṣā''sīd alomikā. Te abruvan tasmai kāmāya alabhāmahai, yathā'syām oṣadhyaś ca vanaspatayaś ca ajāyanta.

[Meaning] Or by that time the earth was barren without any hair growth (vegetations) over it. Let us desire for avi (magnetosphere of the earth), so that vegetations may grow on the earth like hairs on the body.

The above observations indicate the essentiality of Avimedha or harnessing avi energy for the growth of vegetations and life on earth. In the *Yajurveda* (13.50) avi is attributed with the epithet of *ūrṇāyu* (woolen cover) and described as protective skin cover of *paśus* (*tvacaṁ paśūnām*)- *imam ūrṇāyuṁ varuṇasya nābhiṁ tvacaṁ paśūnām.* Here it may also be noted that field lines of intermediate space act as a woolen filter for solar radiation to reach the earth. These field lines, in fact, act as the skin cover of the earth to shield it from solar winds. That is why the avi is attributed with the above epithets.

7

AJĀ : UNBORN ENERGY

In the *Śatapatha Brāhmaṇa* (6.5.1.4), ajā (prakṛti/energy) is described as the form of all animals (particles). It signifies ajā as the unborn energy of the universe. Unborn energy means the energy that has not converted into matter. For this universe to get going energy has to convert into matter. It is the form of all other animals (particles), like aśva (solar energy), gau (geothermal energy) and avi (field energy or energy located in magnetosphere). Ajā in the *Atharvaveda* (9.5.7) is described as agni (energy). The Atharvaveda (9.5.13) further says that Ajā was born from the perturbation of Agni (energy). The *Yajurveda* (13.51) and the *Atharvaveda* (4.14.1) also maintain that ajā was born from a perturbation (śoka) of agni (energy) and he saw agni (energy) first and gods (energy particles) became gods (matter particles) due to aja (energy). This all points out that energy is the source of origin of this visible world. These mantras also point out that ajā is more like a wave form of energy than a particle form of energy. Ajā is described as having one foot (ekapāda) in the *Ṛgveda* (7.35.13) and the *Śatapatha Brāhmaṇa* (8.2.4.1). The ajā is compared to a night in the *Yajurveda*. In the *Yajurveda*

(23.54), ajā is described as piśiṅgilā, meaning devourer of visible matter. The visible universe is born of energy and again consumed into it. Ajā is piśiṅgilā which devours universal objects during dissolution. Ajā also means unborn or sometimes it is called once born (ekaja).

Thus Ajā is the energy and Ajāmedha is nothing else but the process of creation of matter from energy.

Here it may be pointed out that in Agnihotra there is a provision of ājyāhutis. Ājyāhutis are offered with ghee. The ājyāhutis symbolize the ahuti of energy in the yajña of creation. When āhuti of energy is offered in the yajña of creation, matter is formed. Ghee also symbolizes energy.

8

GAU: THE GEOTHERMAL ENERGY

The Vedic seers were well aware of geothermal energy. Śatapatha Brāhmaṇa (14.9.4.19) says that the earth holds energy in her womb. This fact is clarified more vividly in Yajurveda (11.57), as *mātā putraṁ yathopasthe sāgniṁ bibhartu garbha ā* [The earth holds the energy in her womb like the mother her child]. Śatapatha Brāhmaṇa (6.5.1.11) further explains this - *iti yathā mātā putramupasthe bibh'yād evam agniṁ garbhe bibhartviti* [Just as a mother carries the child in her womb, similarly earth holds the energy in her womb]. The same fact has been upheld by Ś.Br. (6.5.5.11) and Tāṇḍya Brāhmaṇa (10.1.1). They maintain that this earth is three-layered: energy - hard crust - vegetation. The energy layer is surrounded by the hard crust which is further covered with vegetation. Taittirīya Brāhmaṇa (3.11.1.17) mentions that energy is located in the earth - *agnir asi pṛthivyāṁ śritaḥ* [Energy has its shelter in earth]. At another place (1.1.3.3) the same fact is mentioned in an allegorical manner as -

agnir devebhyo nilāyata. ākhūrūpaṁ kṛtvā sa pṛthivīṁ prāviśat [Energy absconded from luminary bodies and pierced into the earth like a mouse]. The seer of Śatapatha Br. (6.4.1.2) describes geothermal energy as located in the middle of earth - *pṛthivyā upasthād agniṁ paśavyaṁ*, [I harness energy beneficial to living beings from the middle of the earth]. The above statement points out the harnessing of geothermal energy for various beneficial uses.

Here it is also significant to understand that the Nighaṇṭu reads 'gau' among the names of earth. As such in the Vedas, the intended meaning of gau is earth. In Śrauta Sūtras also gau is intended for earth instead of its apparent meaning 'cow'. Here it is important to know that in all the Vedic rituals, although we find the mention of Puruṣamedha, Aśvamedha, etc., but there is no direct mention of Gomedha available in them. Yes, we come across a ritual named Gavāmayana which is a year-long process. This ritual does not mention the sacrifice of 'cow'. Aitareya Brāhmaṇa (4.17) describes the Gavāmayana ritual. In the opening stanza, it is observed,

gavām ayanena yanti. gāvo vā ādityāḥ. ādityānāmeva tad ayanena yanti. gāvo vai satram āsata.

> [Meaning] Gau is earth and ayana are movements. Gavāmayana symbolizes the movement of the earth (around the sun). Gaus are verily the rays of

the sun. Gavāmayana, therefore the other way round, represents the tropical movements of the sun or āditya on earth. The Uttarāyaṇa (summer solstice) and Dakṣiṇāyana (winter solstice) are two tropical movements of sun or Āditya. Thus Gavāmanayan symbolises Uttarāyaṇa and Dakṣiṇāyana.

Here it is important to know that Gomedha or Gavālambha symbolizes the energy of earth transmitted from its centre that sterilizes the entire global surface thereby enabling us to inhabit it. For want of this geothermal radiation, the earth would have become a desert planet devoid of bio-life.

The 12th mantra of the '*Bhūmi Sūkta*' talks about a perennial source of energy in the womb of the earth. It reads as follows:

yat te madhyam pṛthivī yacca nābhyāṁ yāst urjaḥ tanvaḥ sambhuvuḥ.

tāsu no dhehyabhi naḥ pavasv mātā bhumiḥ putro ahaṁ prithvyāḥ

[Meaning] O Earth! in the midst of your body there is a source of energy, situated exactly at the centre, in your navel. This is your most thematic feature, energizing your entire body. We ought to focus our full attention here only. This is your sanctum sanctorum. The energy transmitted from centre sterilizes the entire global surface thereby enabling us to inhabit it. You are vitalizing and sheltering the entire civilization as the mother feeds and looks after her children.

Interestingly in 1993 American Geophysicist John Marvin Herndon discovered a gigantic self-sustained natural nuclear reactor at the centre of the earth producing 4 terawatts of heat power output to feed the energy requirement of 1343 active volcanoes, over 10000 hot water springs, movement of lithosphere plates, mid-plate earthquakes, hotspots, tsunamis, mountain building and global heat flow value on the surface of the earth.

Figure-3

AFRICA
Gabon
Oklo
The uranium isotopes found at Oklo strongly resemble those in the spent nuclear fuel generated by today's nuclear power plants.

The radioactive remains of a natural nuclear fission reaction that happened 1.7 billion years ago in Gabon, Africa, were held in place by the surrounding geology

Thus the very idea that life on earth is sustained by the sun alone is wrong. The fact is that the earth is equally responsible for the sustenance of life on it.

The planets without their own energy are unable to sustain life even in the presence of the sun. It is the permutation and combination of solar radiation and geothermal energy that generates the atmosphere on earth conducive to sustain bio-life on the earth.

For earthlings there are two basic sources of energy: Firstly, the Self Sustained Natural Nuclear Fission Reactor operating in the core of the Earth and Secondly, the Self-sustained Natural Nuclear Fusion Reactor that in the core of the Sun.

Due to its internally generated energy earth is capable of radiating more energy into space than it receives from the Sun.

The combined effect of solar radiation and geothermal energy drives the atmosphere and oceans into the patterns of everyday wind, water and weather and allows the earth to maintain an average surface temperature of 15^0C.

Various Permutations and combinations of geothermal energy and solar radiation are funding the so-called conventional and non-conventional sources of energy viz. fossil fuels, hydel energy, wind energy, nuclear power reactors, geothermal power plants and solar power plants and so on.

The role of solar radiation in the growth and development of terrestrial plants (through photosynthesis) and those of marine organisms including phytoplankton and zooplanktons is noticed by everybody.

The temperature of the caldera of an erupted volcano reaches up to 1200^0 Centigrade which is about 400^0 Centigrade higher than that of a fossil fuel-fired furnace of an Electric Power Plant. There are 550 active volcanoes and more than 100,000 hot water springs on the earth to provide enough geothermal energy.

Geothermal electricity generation is currently used in 24 countries, while geothermal heating is in use in 70 countries. Estimates of the electricity generating potential of geothermal energy vary from 35 to 2,000 GW. The current worldwide Installed Capacity is 10,715 MW. In recent years, the Indonesian Govt. has announced plans for two 'fast-track' increases in the total capacity of Indonesia's electricity generation network of 10,000 MW each. Geothermal power is considered to be sustainable because the heat extraction is small compared with the earth's heat content. This heat naturally flows to the surface by conduction and is replenished by radioactive decay. The earth's heat content is 10^{31} joules. The estimated electricity generating potential of geothermal energy can readily provide power at rates, more than double humanity's current energy consumption from primary sources.

The emission intensity of existing geothermal electric plants is on average 122 Kg of CO_2 per megawatt-hour (MW-h) of electricity, about one-eighth of a conventional coal-fired plant. As a result, geothermal power has the potential to help mitigate the global warming if widely deployed in place of fossil fuels. Geothermal has minimal land and freshwater requirements. Geothermal plants use 3.5 square kilometres per gigawatt of electrical production versus 32 square kilometres and 12 square kilometres for coal facilities and wind farms respectively. They use 20 litres of freshwater per MW-h versus over 1000 litres per MW-h for nuclear, coal, or oil. Moreover, geothermal power does not rely on variable sources of energy, unlike, e.g. wind or solar. Its capacity factor can be quite large - up to 96% has been demonstrated. The global average was 73% in 2005.

9

PURĪṢA AGNI: THE BIO-ENERGY OF EARTH

The *Kapiṣṭhala Kaṭha Saṁhitā* (35.3) maintains that energy is also located on earth in the form of Purīṣa (excretion i.e.fossil fuel)- *ye agnayaḥ purīṣiṇa āviṣṭā pṛthivim anu* (i.e. these energies have entered into the earth in the form of fossil fuels). At another place, the same fact has been substantiated as *māteva putraṁ pṛthivī purīṣyam agniṁ sve yonāvubhārukhā* [Just as mother delivers a child, similarly the earth delivers energy in the form of fossil fuels]. RV (1.163.1), describes fossil fuels as one of the sources of energy.

yad akrandaḥ prathamaṁ jāyamāna udyant samudrād uta vā purīṣāt

śyenasya pakṣā hariṇasya bāhū upastutyaṁ mahi jātaṁ te arvan

[Meaning] Your great birth, O glorified energy is to be decried; whether first springing from the sea bottom or from excrement (faeces) of earth. You have the wings of the falcon and the limbs of the deer.

The above mantra notes that energy also originates from the sea bottom and land from the faeces or excrement of the earth. This, the other way round, points out to the generation of energy in anaerobic decomposition of remains of buried dead organisms (in the earth's crust and the sea bottom) and their further chemical alteration into fossil fuels (like coal, methane, natural gas and petroleum). The energy thus generated is called Purişan energy. Purişa here symbolises excrement or faeces of earth. The combustion of fossil fuels releases energy which is second-hand/ recycled forms of solar and geothermal Energies. That is why, Genius Mother Nature has deliberately dumped these rotten, poisonous products (fossil fuels) of solar radiation and geothermal energy, many kilometers deep into the continental and oceanic crusts, The harnessing energy from feces of the earth is decried, however, glorified it may be, as the consumption of these fossil fuels for purpose of energy will ultimately vitiate the biosphere of earth.

CONCLUSION

In view of the above discussion, it can unhesitatingly be maintained that the Vedic seers divided energy into two categories -mechanical and psychic- according to their view of the origin of the universe from twins i.e. Puruşa and Prakṛti. Puruşa is represented by psychic energy and Prakṛti by mechanical energy. They were very conscious about the healthy and clean environment and never sanctioned any act that may lead to pollution of the atmosphere. They stressed the need to harness the solar and geothermal energy for use by earthlings instead of fossil fuel energy.

REFERENCES

1. *Ṛgveda Saṁhitā*, ed. Ravi Prakash Arya, 2nd Edition, Parimal Publication, Delhi, 2001.

2. *Yajurveda Saṁhitā*, ed. Ravi Prakash Arya, 2nd Edition, Parimal Publication, Delhi, 2002.

3. *Sāmaveda Saṁhitā*, ed. Ravi Prakash Arya, 2nd Edition, Parimal Publication, Delhi, 2001.

4. *Atharvaveda*: Translated into Hindi by Kṣema Karaṇa Trivedī, Dayananda Sansthan, New Delhi.

5. *Atharvaveda*: Edited by Raghvir, S. V. Granthamāla, Lahore, 1936.

6. *Śatapatha Brāhmaṇa*: Translated into English by Eggeling, J. Oxford, Clarendon Press.

7. *Caraka Saṁhitā*, ed. Ram Karan Sharma & Vaidya Bhagwan Dash, Chokhamba Sanskrit Series Office, Varanasi, 2011.

8. Taittirīya Saṅhitā: Edited by Shripada Sharma, Aundh, 1945.

9. *Taittirīya Brāhmaṇa*: With the commentary of Bhaṭṭabhāskara Miśra edited by Mahavir Sastrin, Mysore, 1908-21.

10. *Śāṅkhāyana Āraṇyaka*: Edited by Swami

Shridhara Shastrin, Anandashram, Poona, 1922.

11. *Kāṭhaka Saṅhitā*: Edited by Shripada Sharma, Aundh, 1943.

12. *Maitrāyaṇī Saṅhitā*: Edited by Shripada Sharma, Aundh, 1943.

13. *Gopatha Brāhmaṇa*: Translated in Hindi by KṣemakaraṇaTrivedi, Prayāga, 1920.

14. *Jaiminīya Brāhmaṇa*: Edited by Raghvir, Lokesh Chandra, SVS Nagpur, 1954.

15. *Kaṭha Saṁhitā*: Edited by Shripada Sharma, Aundh, 1942.

16. *Aitareya and Kauṣitaki Brāhmaṇas of Ṛgveda*: Edited by Keith, A.B., Delhi, Motilal, 1971.

www.ingramcontent.com/pod-product-compliance
Lightning Source LLC
Chambersburg PA
CBHW070751220326
41520CB00053B/3885